Computer Intelligence Engineers

Anne Taylor

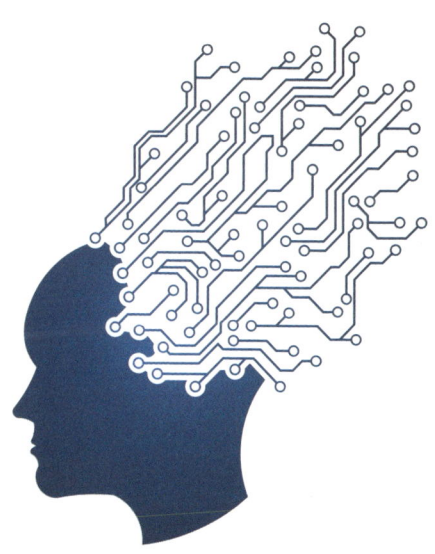

Series Editor **Casey Malarcher**

Level 2 - ❹

Computer Intelligence Engineers

Anne Taylor

© 2018 Seed Learning, Inc.

Series Editor: Casey Malarcher
Acquisitions Editor: Kelly Daniels
Copy Editor: Liana Robinson
Cover/Interior Design: Highline Studio

ISBN: 978-1-943980-41-3

10 9 8 7 6 5 4 3 2 1
22 21 20 19 18

Photo Credits

All photos are © Shutterstock, Inc.

Contents

What Is a Smart Computer?

We can think and learn using our brains. Computers can't think. They don't have brains. But one day they will. Smart computers will think. And they will learn new things.

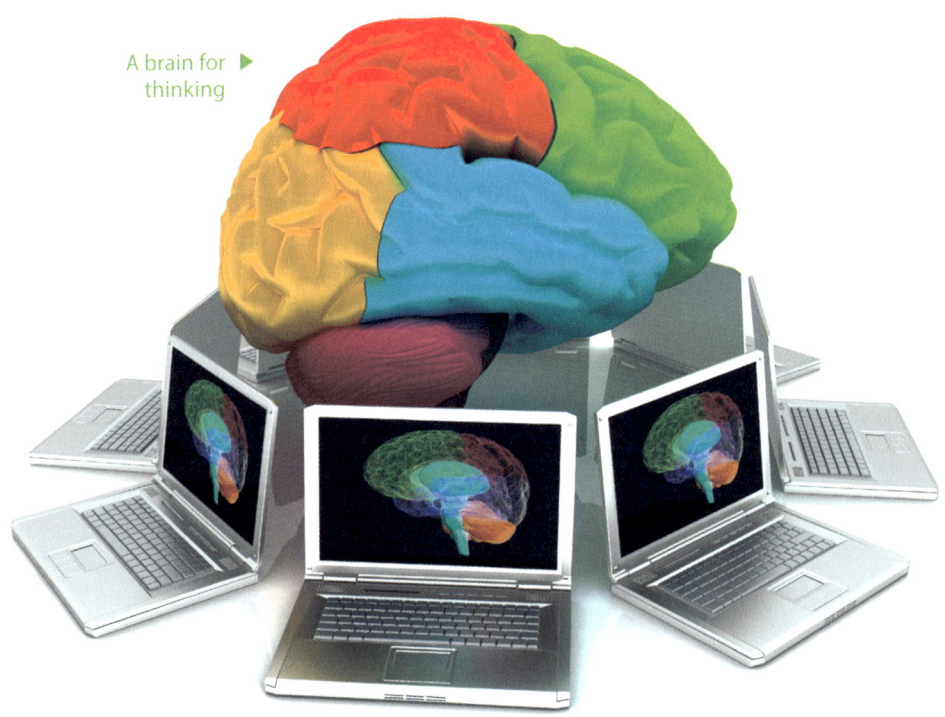

A brain for ▶ thinking

Who will make a computer's brain?

A computer intelligence engineer!

Working on a computer

What do computer intelligence engineers do?

Computer intelligence engineers make smart computers. Well, they are trying to! They want to make computers that can think and understand. It's a difficult thing to do. So they are not done yet.

Developing new computer technology

A company called IBM made a smart computer called Watson. Watson went on a TV game show. It answered lots of questions and won the game. Even though Watson can do a lot, engineers want to make a better computer. They are trying very hard.

The TV game show that Watson was on

A university

Smart computers will mimic the way the human brain works. They will understand language and recognize objects. They will perform the same kinds of tasks as humans do.

Researchers at some universities are working together with a computer company. They are making a super-computer. This super-computer will perform like a real brain.

A computer chip

How will it perform like a real brain? The university researchers are working on how to make a new chip. The chip will mimic the way our brains send information to other parts of our bodies.

Researchers at a university in California are preparing a kind of video game to help train the computer. When it is ready, it will be used to teach the computer how to learn and decide what to do.

Kids playing a video game

Why Do We Need Smart Computers?

Smart computers will think like people. But they will think more quickly.

They will also remember more information than humans. So they can find answers to very difficult problems for humans.

◀ A woman with an idea

Doctors could use smart computers in hospitals. A smart computer could recognize a patient's face. It could remember information about the patient. It could even know if a patient is really sick or not. It could tell the doctor how to treat the patient. Smart computers won't replace doctors, but they will help doctors work more quickly. They will help doctors be better at their jobs.

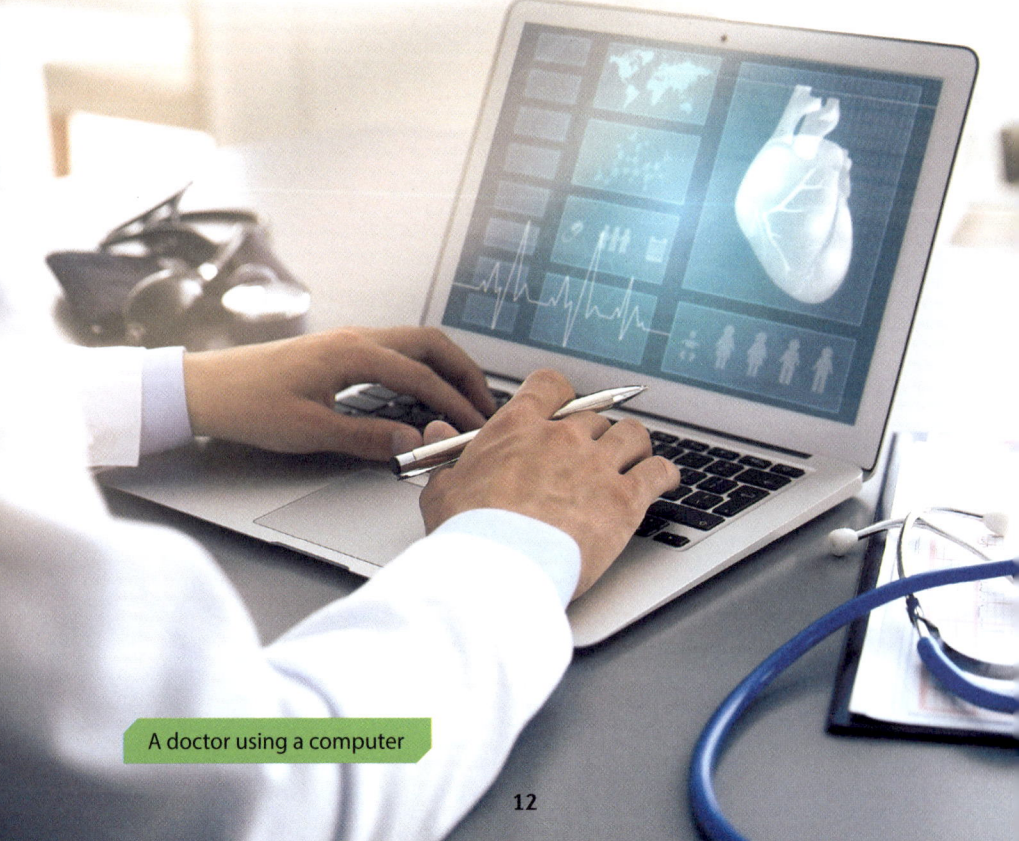

A doctor using a computer

A blind woman

Smart computers will also be able to help blind people. In Japan, there is a famous computer scientist with IBM who is blind. She is working to develop technology to help other blind people. She wants to make a computer that can recognize faces. It will tell her when her friends are near, so she can say hello.

Walking past a store window

When a blind person passes a store window, it will tell her or him what is for sale.

Smart computers will tell blind people how their friends are feeling. The computers will make it easier for them to feel closer to their friends. Smart computers will make life much easier for blind people.

Smart computers could help people feel better. Sometimes, people have very bad experiences in their lives. For example, they have a bad accident, or they are soldiers who go to war. After these people return home, life is difficult for them. Many things make them feel sad or upset.

A man who ▶
feels sad

A hospital for soldiers introduced Watson to some patients. Watson can answer complex questions quickly in natural language. Maybe someday, Watson will tell patients ways to feel better. In the future, people could call a computer when they are feeling bad.

◄ A hospital patient recovering from a bad accident

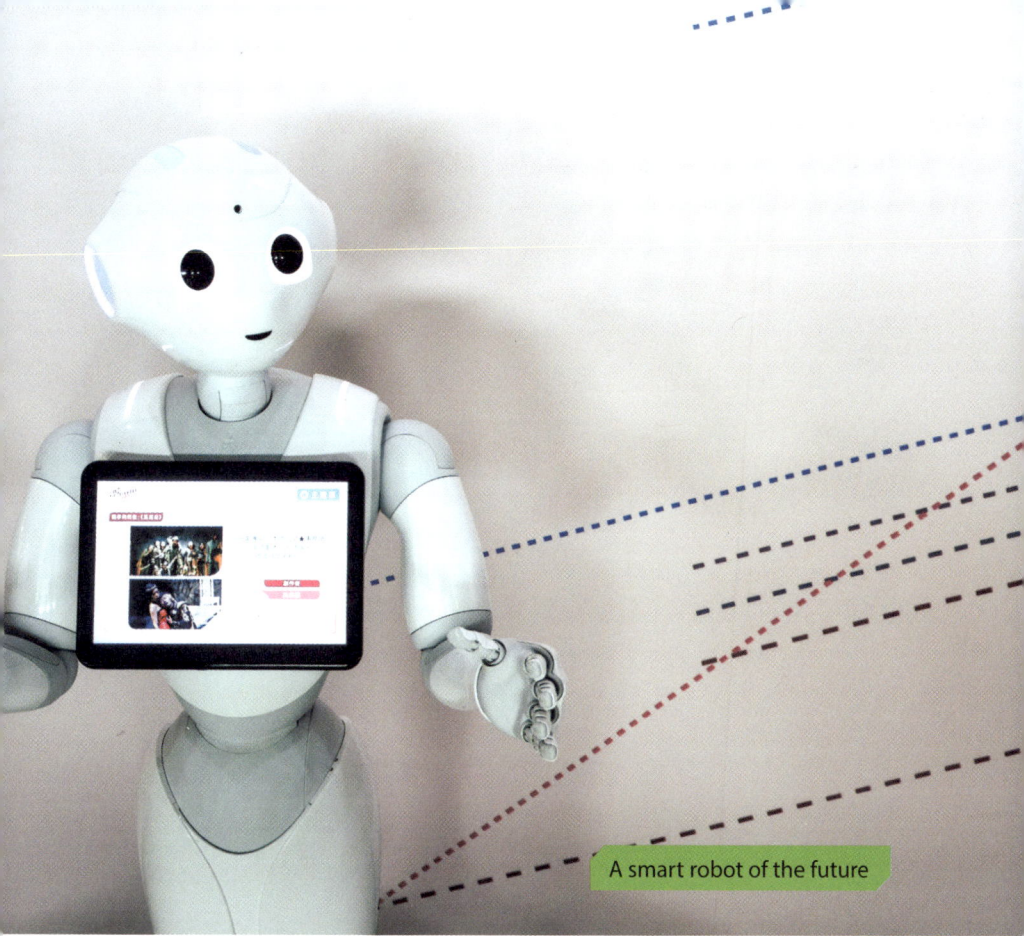

A smart robot of the future

Smart computers can also be used inside robots. One hotel recently introduced a robot called Connie. Using a smart computer, Connie can answer questions. Visitors can ask about the hotel and the local area. Connie can understand questions and recommend restaurants.

How to Be a Computer Intelligence Engineer?

A computer intelligence engineer must be a good thinker. Making a new kind of computer is difficult. Smart computers need smart makers.

Thinking about ▶
making a smart
computer

Saving data

The engineers also need to be organized. They need to save a lot of data. Sometimes they must save it for years. Of course, they must be able to find it later!

Asking questions

Computer intelligence engineers must like asking questions. And they must be good at finding the answers. A lot of thinking is needed to do this job. Making smart computers is hard. So computer intelligence engineers discuss and share ideas. They must be good at working with others in teams.

Studying the brain

Of course, computer intelligence engineers must know a lot about computers. And they must know a lot about people's brains, too.

What Do Computer Intelligence Engineers Study?

You will need a degree in computer science. You should also study some math and biology. You need to study biology so that you can understand the brain.

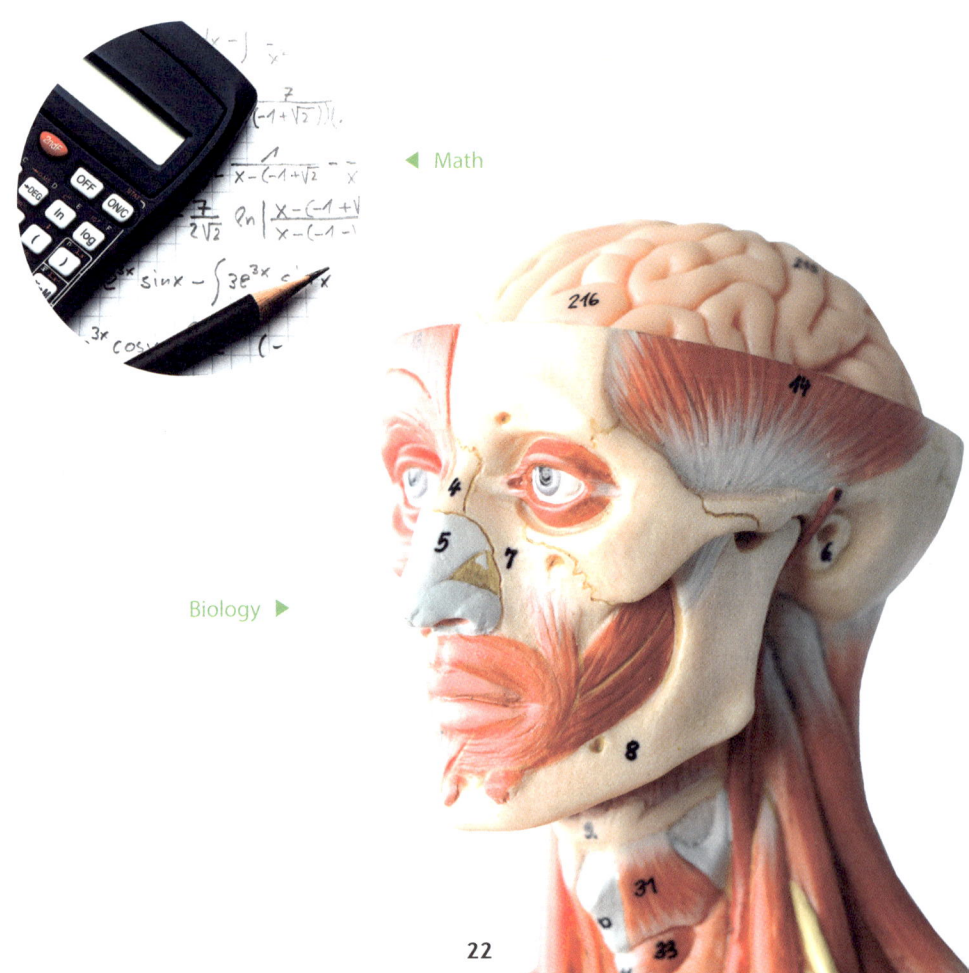

◀ Math

Biology ▶

A master's degree is best if you want a good job. If you get a Ph.D., you could try to make a new discovery about computers. The idea of smart computers is still new, so there is a lot more to learn.

Getting a university degree

Where Are the Jobs?

Some computer intelligence engineers work in universities. They do research or teach students how to make smart computers.

Working at a university

Computer code

Other computer engineers work for big companies. They help make new smart computers. Or they figure out new computer codes. They try to think of ways of making better computers.

In the Future?

In the future, you will be able to talk to your computer. You will be able to talk to your phone, car, or even your house. Your computer will understand you. It will know how you are feeling. It will talk to you like a friend.

Future communication

Do you want to make smart computers that understand people? Then computer intelligence engineering might be for you!

Future computer intelligence engineers

Comprehension Questions

1. Smart computers will . . .
   (a) cook and clean.
   (b) go to university.
   (c) help blind people see again.
   (d) think and understand.

2. What should a computer intelligence engineer study?
   (a) English and math
   (b) Computer science and biology
   (c) Math and business
   (d) None of the above

3. Which is NOT something a smart computer could do?
   (a) Run fast
   (b) Help doctors
   (c) Recognize faces
   (d) Help blind people

4. Where do computer intelligence engineers work?
   (a) On TV shows
   (b) In taxis
   (c) At universities
   (d) All of the above

5. Our ideas for smart computers are still new, so . . .
   (a) we know everything about them.
   (b) there is a lot more to learn about them.
   (c) they are big.
   (d) they are very common.

Glossary

- **biology** (n.) the study of living things

- **data** (n.) information, especially numbers and facts; information stored in a computer

- **degree** (n.) the qualification obtained by students who successfully complete a university or college course

- **engineer** (n.) a person who thinks of new ways to build things or solve problems

- **intelligence** (n.) the ability to learn, think, and understand

- **mimic** (v.) to copy; to act like someone or something else

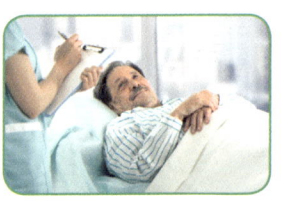

- **patient** (n.) a sick person who visits a doctor for help

- **recognize** (v.) to know someone or something, usually because you've seen them before

- **researcher** (n.) a person who studies something in detail to get new information

- **software** (n.) the programs, etc. used to operate a computer

- **task** (n.) a piece of work to be done

- **technology** (n.) the study and knowledge of scientific discoveries

Notes

Here are some of the important names in the field of intelligent computing (also known as cognitive computing). Readers may enjoy researching these people and places to learn more about this field.

Dharmendra S. Modha is a computer researcher who is leading a team trying to make a computer that can act like a human brain. In 2009, he said his team had managed to make a computer mimic a cat's brain.

Almaden Research Center is a research lab in San Jose, California. It is where many of the world's top scientists investigate computer science.

Chiecko Asakawa is a computer scientist with IBM in Tokyo who developed computer programs to help blind people. She has received many awards for her research, and she can be seen in TED talks.

Watson is a question-answering computer system able to understand natural language. It was developed at the computer company IBM. Watson was named after IBM's first CEO, Thomas J. Watson.

List of Books

LEVEL 1

❶ Robotics Engineers

❷ Cyber Security Experts

❸ Medical Scientists

❹ Social Media Managers

❺ Asset Managers

LEVEL 2

❶ Drone Pilots

❷ App Developers

❸ Wearable Technology Creators

❹ Computer Intelligence Engineers

❺ Digital Modelers

LEVEL 3

❶ IoT Marketing Specialists

❷ Space Pilots

❸ Water Harvesters

❹ Genetic Counselors

❺ Data Miners

LEVEL 4

❶ Database Administrators

❷ Nanotechnology Research Scientists

❸ Quantum Computer Scientists

❹ Agricultural Engineers

❺ Intellectual Property Lawyers

"The future of the economy is in STEM. That's where the jobs of tomorrow will be."

James Brown (Executive Director of the STEM Education Coalition in Washington, D.C.)

Data from the US Bureau of Labor Statistics (BLS) support that assertion. Employment in occupations related to STEM—science, technology, engineering, and mathematics—is projected to grow to more than 9 million by 2022 [in the US alone] . . . Overall, STEM occupations are projected to grow faster than the average for all occupations.

from *STEM 101: Intro to Tomorrow's Jobs* **US Bureau of Labor Statistics**